Introduction to Morse Code

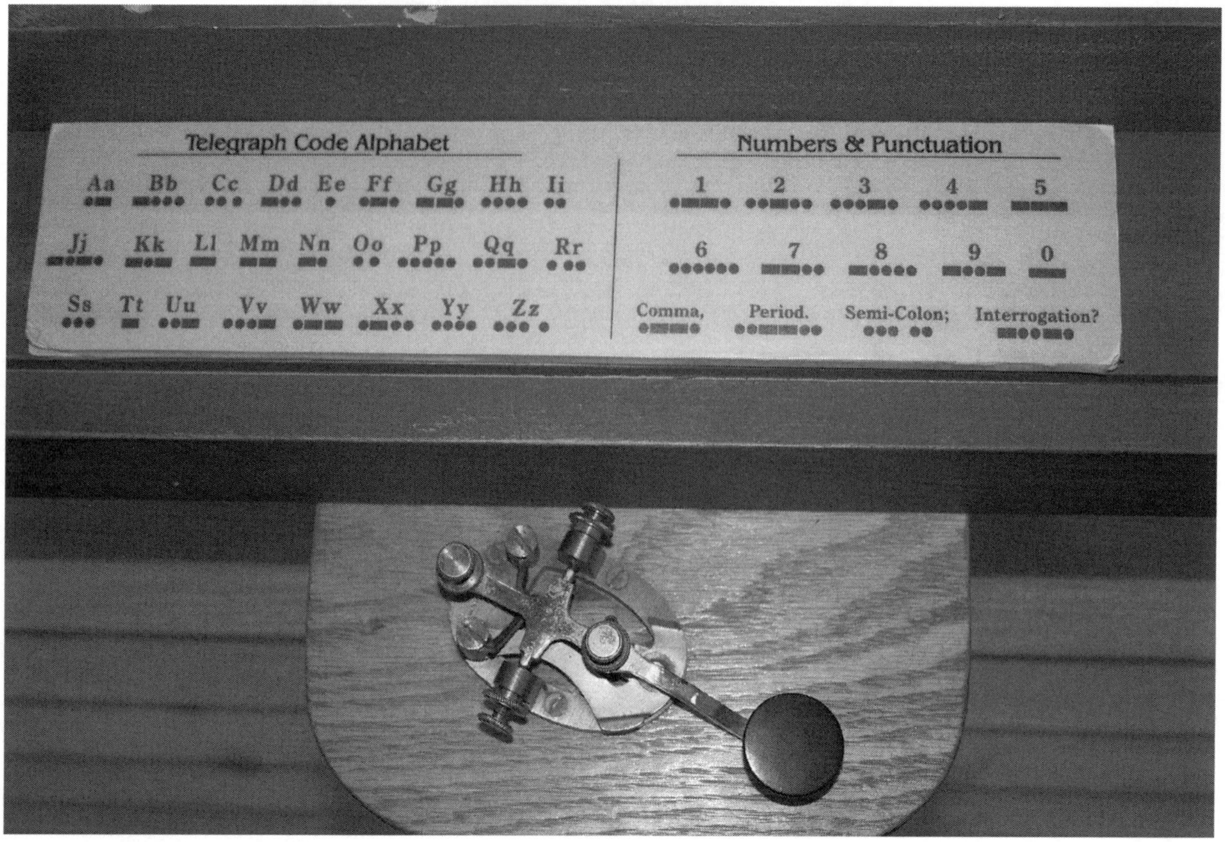

The Significance of Morse Code in Communication History

Morse code, with its rhythmic arrangement of dots and dashes, stands as a testament to the indomitable human spirit of innovation and the relentless pursuit of efficient communication. In the grand tapestry of communication history, Morse code occupies a pivotal role, leaving an enduring mark that reverberates through time.

Before the digital age, when instant messaging and global connectivity were mere dreams, Morse code was the conduit of choice for transmitting messages across vast distances. Its simplicity in design, yet profound implications, elevated it to an indispensable tool for countless endeavors, shaping the course of events in ways that continue to influence our world.

At its heart, Morse code represents a bridge between the old and the new, a connection from an era when the telegraph's rhythmic clicks heralded the dawn of a new era of information exchange. This system of encoding text into audible and visible signals revolutionized communication in ways that were unimaginable before its invention.

THE ULTIMATE

MORSE CODE HANDBOOK

JONATHAN HARRISON

Maritime communications, for instance, were revolutionized by Morse code. Ships once isolated at sea could now relay vital information, enabling safer navigation and timely responses to emergencies. The rhythmic dance of dots and dashes transcended language barriers, becoming a universal language of the high seas.

The historical importance of Morse code extends beyond the sea, reaching into the heart of land-based communication networks. The telegraph lines crisscrossing continents, connecting cities and towns, facilitated rapid information transfer, making distant events feel closer than ever before. News that once took days or weeks to travel could now traverse vast distances in the blink of an eye, transforming journalism, commerce, and diplomacy.

In times of conflict, Morse code emerged as an invaluable asset. Its use in military communication allowed commanders to coordinate strategies with precision, leading to more effective operations on the battlefield. It became a lifeline for soldiers, providing a means to request reinforcements, relay critical information, and maintain the morale of those far from home.

As we explore Morse code, we're not just learning a communication system; we're unlocking a treasure trove of historical significance. We're connecting with the minds and moments that shaped our past, acknowledging the ingenuity that paved the way for our present interconnected world. Morse code, with its dots and dashes, reminds us that the quest for efficient and effective communication is a timeless pursuit—one that echoes through history and propels us toward a future where the art of connection continues to evolve.

Why Learning Morse Code Is Still Relevant Today

In a world dominated by smartphones, instant messaging, and high-speed internet, it might seem curious to embark on a journey to learn a communication system as seemingly antiquated as Morse code. However, as we delve into the reasons why Morse code remains relevant in the modern age, you'll discover that this timeless skill offers far more than mere historical intrigue—it opens doors to a range of practical and profound benefits.

1. Resilient Communication: In an era where digital networks can falter during emergencies, Morse code shines as a reliable backup. Its simplicity allows for communication even in situations where modern technology fails. For adventurers, hikers, sailors, and those in remote areas, Morse code can be a lifeline, enabling them to signal for help when other means are unavailable.

2. Cognitive Enhancement: Learning Morse code is a mental workout that sharpens memory, attention, and cognitive skills. Deciphering the rhythmic patterns of dots and dashes hones your ability to focus, improves listening skills, and boosts overall mental agility.

3. Personal Achievement: Mastering Morse code brings a sense of accomplishment that comes from acquiring a unique skill. It's a fascinating hobby, a talking point in social settings, and a badge of intellectual curiosity that sets you apart from the crowd.

4. Historical and Cultural Appreciation: Delving into Morse code opens a window into the past, allowing you to connect with the rich history of communication technology. It's a nod to the pioneers who paved the way for our modern information age. Moreover, Morse code's influence extends beyond technology, as it has been a part of military, maritime, and cultural history.

5. Amateur Radio and Community: Morse code is a fundamental part of amateur radio (ham radio) culture. Learning Morse code enables you to join this global community of enthusiasts who engage in long-distance communication, public service, and experimentation. It's an opportunity to connect with like-minded individuals around the world.

6. Intellectual Exploration: Understanding Morse code is a gateway to understanding the basics of encoding, decoding, and data transmission. It provides a foundation for grasping more complex communication systems and technologies.

As you embark on this journey to learn Morse code, you're not just acquiring a historical curiosity; you're immersing yourself in a skill that offers practical applications, intellectual growth, and a deeper appreciation for the evolution of human communication. Morse code, with its rhythmic dance of dots and dashes, continues to find relevance in our modern world, reminding us that sometimes, the most enduring lessons are found in the echoes of the past.

Overview of the book's structure and learning approach

This book is designed as a comprehensive guide to help you master Morse code, from its origins and fundamental principles to practical application. We've carefully crafted a structured approach that combines historical insights, theory, practical exercises, and engaging activities to ensure a rewarding learning experience. Here's a preview of what to expect from each section:

1. Introduction to Morse Code: We'll dive into the historical significance of Morse code, exploring its impact on communication history, from the days of telegraphy to its modern-day relevance.

2. The History of Morse Code: This chapter delves deeper into the fascinating journey of how Morse code came to be, showcasing its pivotal role in transforming long-distance communication.

3. Understanding the Morse Code Alphabet: Get ready to familiarize yourself with the dots and dashes that make up the Morse code alphabet. We'll cover characters, abbreviations, and prosigns, laying the foundation for effective Morse code communication.

4. How the Morse Code System Works: Discover the logic behind Morse code encoding, timing, and transmission. We'll demystify the core principles that make Morse code an efficient and versatile communication tool.

5. Practice Makes Perfect: Letters and Numbers: In this hands-on section, you'll learn the Morse code alphabet inside out. Engage in exercises that hone your transmission and reception skills, ensuring you become proficient at deciphering messages.

6. Advanced Morse Code Techniques: Explore the finer nuances of Morse code, including punctuation, prosigns, and other special sequences. Unlock tips for efficient communication and delve into advanced applications.

7. Exploring Morse Code Applications: Learn how Morse code continues to find relevance in various contexts, from emergency situations to amateur radio operations. Discover its role in modern technology and communication systems.

8. Putting It All Together: Apply your newfound knowledge to real-life scenarios, complex message encoding, and decoding challenges.

9. Practice Exercises:
Build confidence in your Morse code skills through hands-on practice.

Throughout the book, we encourage a holistic learning approach, combining theory with hands-on practice. Each chapter ends with exercises to reinforce what you've learned, helping you gradually build confidence and proficiency in Morse code. Whether you're a beginner exploring the world of communication or someone looking to enhance your cognitive skills, this book provides a structured pathway to unlock the beauty and practicality of Morse code. Get ready to embrace the rhythmic dance of dots and dashes and embark on a rewarding journey of learning and discovery!

The History of Morse Code

Samuel Morse and the Invention of the Telegraph

The story of Morse code begins with a brilliant and determined inventor named Samuel Finley Breese Morse. Born on April 27, 1791, in Charlestown, Massachusetts, Morse would grow up to become a key figure in the history of communication, forever changing the way the world shared information across great distances.

Before delving into Morse's groundbreaking invention, it's important to recognize that his early career was marked by artistic pursuits. Morse was a talented painter, known for his historical and portrait works. He studied art in the United States and Europe, honing his skills and gaining recognition for his contributions to the world of visual arts.

However, it was during a voyage across the Atlantic in 1832 that Morse's life took a transformative turn. News arrived via ship that his wife, Lucretia, was critically ill. By the time Morse reached New Haven, Connecticut, he received the devastating news that she had passed away. The tragedy of not being by her side during her final moments due to the sluggishness of communication deeply impacted Morse and ignited a desire to find a faster and more efficient means of long-distance communication.

This desire led Morse to conceive the idea of the electric telegraph, a revolutionary concept that would enable messages to be transmitted instantaneously over long distances using electrical signals. Morse, driven by his passion for innovation, began working on the development of this groundbreaking technology. Together with the support and expertise of his partner, Alfred Vail, Morse refined the design and code that would later bear his name.

In 1838, Morse, Vail, and a small team completed a working prototype of the telegraph system, which used a system of dots and dashes to represent letters and numbers. The simplicity of Morse code allowed for clear and efficient transmission of messages, making it a pivotal component of the telegraph's success.

In 1844, Morse achieved a historic milestone when he sent the first long-distance telegraph message from Washington, D.C., to Baltimore, Maryland, with the iconic phrase "What hath God wrought." This momentous event marked the birth of instantaneous communication over vast distances, transforming the way society operated. It connected cities, facilitated commerce, and changed the dynamics of news dissemination, forever altering the course of history.

Morse's invention of the telegraph, powered by the ingenious Morse code, had a profound impact on the world. It laid the foundation for modern telecommunications, paving the way for the global interconnectedness we enjoy today. Samuel Morse's legacy as an artist, inventor, and visionary lives on, and his contribution to the realm of communication remains one of the most influential innovations in human history.

The role of Morse code in maritime communication

In the vast expanse of the open sea, where distances are measured not in miles but in the endless horizon, reliable communication has always been a maritime imperative. Morse code emerged as a maritime lifeline, connecting ships to distant shores and ensuring the safety and efficiency of seafaring operations. Let's explore the vital role that Morse code played in maritime communication, solidifying its place as an indispensable tool at sea.

1. Navigational Assistance: The sea can be unforgiving, with unpredictable weather patterns and treacherous obstacles. Morse code became a crucial means of transmitting navigational information, ensuring that ships could avoid hazards, adjust courses, and safely navigate through challenging waters. The ability to receive timely updates on changing conditions and navigation instructions was essential for the safety of crew and cargo.

2. Ship-to-Shore Communication: Before the era of satellite communication, ships relied on Morse code to maintain contact with shore-based stations. Whether it was relaying cargo manifests, coordinating port entries, or simply staying in touch with loved ones, Morse code facilitated the exchange of critical information between ships and land.

3. Emergency Signaling: In times of distress, when immediate assistance was required, Morse code provided a universal language for signaling SOS (···---···), the internationally recognized distress call. Ships in peril could transmit this distress signal using Morse code, alerting nearby vessels or coastal stations to their urgent need for help.

4. Maritime Codes and Conventions: Morse code formed the foundation of maritime signaling systems, enabling ships to communicate using visual and auditory signals. Semaphore flags, signal lamps, and whistle codes all relied on the principles of Morse code to convey messages, ensuring that ships could communicate even when radio or telegraph equipment was unavailable or compromised.

5. Military and Defense: In times of conflict, Morse code played a pivotal role in maritime defense. Naval vessels used Morse code to transmit vital tactical information, coordinate fleet movements, and maintain operational secrecy. The ability to communicate covertly using Morse code was a strategic advantage during wartime.

6. Connecting the World: The introduction of telegraph cables under the ocean further extended the reach of Morse code, enabling transcontinental communication. Telegraph cables laid across the ocean floor connected continents, and Morse code was the language of choice for transmitting messages across these undersea wires, bridging the gap between distant shores.

As you embark on your journey to learn Morse code, remember that it was the maritime realm that first recognized the unparalleled value of this communication system. The rhythmic sequence of dots and dashes echoed across the waves, forming a vital connection between ships, harbors, and distant shores. The legacy of Morse code in maritime communication is a testament to its reliability, adaptability, and enduring importance in the annals of maritime history.

Morse code's impact on military and espionage

In the world of military strategy and clandestine operations, where information can be a matter of life and death, Morse code emerged as a powerful tool that revolutionized communication on the battlefield and in the shadowy realm of espionage. Its impact was profound, shaping the course of military history and becoming an essential part of covert activities. Let's explore how Morse code became a secret language for the military and a crucial instrument in the world of espionage.

1. Rapid Communication: In the heat of battle, swift and efficient communication is paramount. Morse code provided a method of transmitting critical messages quickly and reliably across the chaos of the battlefield. Whether it was coordinating troop movements, relaying tactical instructions, or sending urgent requests for reinforcements, Morse code allowed military commanders to maintain situational awareness and respond with precision.

2. Secure Communication: Morse code offered a level of security that traditional spoken communication could not guarantee. Encoded messages in Morse code could be transmitted over telegraph lines or via signal lamps, reducing the risk of interception and unauthorized access. This made Morse code a preferred means of conveying classified information and maintaining operational secrecy.

3. Covert Signaling: Espionage and covert operations often rely on discreet signaling to convey information without alerting the enemy. Morse code provided a covert language that spies and operatives could use to transmit critical intelligence, without revealing their messages to prying eyes or ears. Its simplicity made it easier to memorize and transmit discreetly.

4. Resistance Movements: In occupied territories during wartime, Morse code became a vital means of communication for resistance movements. Underground networks and partisan groups used Morse code to coordinate activities, share information about enemy movements, and plan covert operations against occupying forces. This allowed resistance fighters to maintain coordination while minimizing the risk of detection.

5. Radio Communication: The advent of radio technology further expanded the role of Morse code in military and espionage operations. Morse code became the language of radio operators, enabling long-distance communication even in situations where voice transmission was impractical or compromised. During World War II, for example, Morse code was widely used by radio operators on both sides of the conflict.

Understanding the Morse Code Alphabet

A	• —	N	— •
B	— • • •	O	— — —
C	— • — •	P	• — — •
D	— • •	Q	— — • —
E	•	R	• — •
F	• • — •	S	• • •
G	— — •	T	—
H	• • • •	U	• • —
I	• •	V	• • • —
J	• — — —	W	• — —
K	— • —	X	— • • —
L	• — • •	Y	— • — —
M	— —	Z	— — • •

The Basics of Dots and Dashes

At the heart of Morse code lies a simple yet ingenious way of representing letters, numbers, and symbols using two fundamental elements: dots and dashes. These elements serve as the building blocks of Morse code, forming the foundation for clear and efficient communication. Below, we'll explore the significance of dots and dashes, providing a visual understanding that underscores their role in this timeless communication system.

Dots (•):

The dot, represented as ·, is the shortest element in Morse code. It's a quick and distinct signal, serving as the basis for encoding letters and numbers. Dots are used to create the visual and auditory distinctions that define each character in Morse code. A single dot represents a basic unit of time, and all other elements are constructed based on this unit.

Dashes (—):

The dash, represented as —, is a longer signal compared to the dot. It's the counterpart to the dot, providing contrast and creating the unique patterns that make each Morse code character recognizable. Dashes are typically three times the duration of a dot. The combination of dots and dashes enables Morse code to encode a wide range of symbols, from the letters of the alphabet to numerals and special characters.

As you become more familiar with Morse code, you'll discover that each letter, numeral, and symbol is composed of a specific arrangement of dots and dashes. By mastering these basic elements and their combinations, you'll gain the ability to encode and decode messages in Morse code. This foundational knowledge forms the starting point for your journey to unlock the full potential of this versatile communication system.

Keep in mind that Morse code's simplicity is one of its strengths. With practice, you'll quickly become adept at recognizing the patterns of dots and dashes that form the Morse code representations of various characters. As you progress through this book, you'll build upon this understanding, becoming fluent in the language of dots and dashes, and mastering the art of Morse code communication.

Introduction to Morse Code Characters and Their Representations

Morse code is a clever encoding system that uses combinations of dots (·) and dashes (—) to represent letters, numbers, and even some punctuation marks. Each character has a unique Morse code sequence, allowing messages to be transmitted using this concise and efficient method. In this section, we'll introduce you to the Morse code representations for the letters of the English alphabet, numbers, and a few common punctuation marks. The visual representations provided here are in text form, but you can add appropriate diagrams to enhance your learning experience.

Letters of the English Alphabet:

Below, you'll find the Morse code representations for the letters of the English alphabet. Pay attention to the patterns of dots and dashes that form each letter's

Morse code equivalent.

A: • —
B: — • • •
C: — • — •
D: — • •
E: •
F: • • — •
G: — — •
H: • • • •
I: • •
J: • — — —
K: — • —
L: • — • •
M: — —
N: — •
O: — — —
P: • — — •
Q: — — • —
R: • — •
S: • • •
T: —
U: • • —
V: • • • —
W: • — —
X: — • • —
Y: — • — —
Z: — — • •

Numbers:

Morse code also includes representations for numbers. These numeric characters are essential for transmitting numerical information effectively.

0: — — — — —
1: • — — — —
2: • • — — —
3: • • • — —
4: • • • • —
5: • • • • •
6: — • • • •
7: — — • • •
8: — — — • •
9: — — — — •

Common Punctuation Marks:

While Morse code primarily focuses on letters and numbers, it also includes representations for some common punctuation marks that allow for more comprehensive communication.

Period (.) : • — • — • —
Comma (,) : — — • • — —
Question Mark (?) : • • — — • •

As you study these Morse code representations, keep in mind that each character's sequence of dots and dashes is unique, allowing for precise communication using this system. Practice decoding and encoding messages with these Morse code characters, and you'll soon become proficient in this time-tested method of transmitting information through the rhythmic dance of dots and dashes. Feel free to add diagrams to visually reinforce your understanding.

Commonly Used Morse Code Abbreviations and Prosigns

In addition to representing individual letters, numbers, and punctuation marks, Morse code features a set of abbreviations and prosigns that enhance its efficiency and practicality in various communication scenarios. These specialized symbols and sequences allow Morse code users to convey common phrases, convey specific meanings, and streamline transmissions. Here, we'll introduce you to some of the commonly used Morse code abbreviations and prosigns that play a crucial role in effective Morse code communication.

1. AR (End of Message):
AR, or "End of Message," is used to indicate the conclusion of a message or transmission. It serves as a clear signal that the current communication is complete, allowing recipients to prepare for the next message.

2. AS (Wait):
AS, or "Wait," signals that the sender is pausing temporarily to allow the recipient to catch up or respond. It's a valuable prosign for ensuring that messages are received and acknowledged appropriately.

3. KN (Invitation for a Specific Station to Transmit):
KN, or "Invitation for a Specific Station to Transmit," is used to direct a specific station to respond or take their turn in a conversation. It's a way of managing communication flow in a multi-party conversation.

4. BT (Spacing):
BT, or "Spacing," is used to introduce extra spacing between words or phrases. It helps improve message clarity by preventing characters from running together, especially in longer messages.

5. SK (End of Communication):
SK, or "End of Communication," is a prosign indicating the conclusion of the entire communication or conversation. It's a more comprehensive version of "End of Message," often used when the communication session is ending.

6. SOS (Distress Signal):
SOS, with its distinctive pattern of three short signals, three long signals, and three short signals (· · · — — — · · ·), is recognized worldwide as the universal distress signal.

It is one of the most critical Morse code sequences, indicating that urgent assistance is required.

7. CQ (Calling All Stations):
CQ is a general call inviting any station that hears the signal to respond. It's often used in amateur radio operations to initiate contact with other stations, especially in a broadcast-like manner.

8. DE (From):
DE, or "From," is used to indicate the sender of a message. It's commonly used at the beginning of transmissions to identify the originating station or operator.

By incorporating these commonly used abbreviations and prosigns into your Morse code communication, you'll enhance your ability to convey messages more efficiently, ensure proper communication etiquette, and navigate various communication scenarios effectively. Familiarize yourself with these symbols and their meanings to become a skilled Morse code communicator.

How the Morse Code System Works

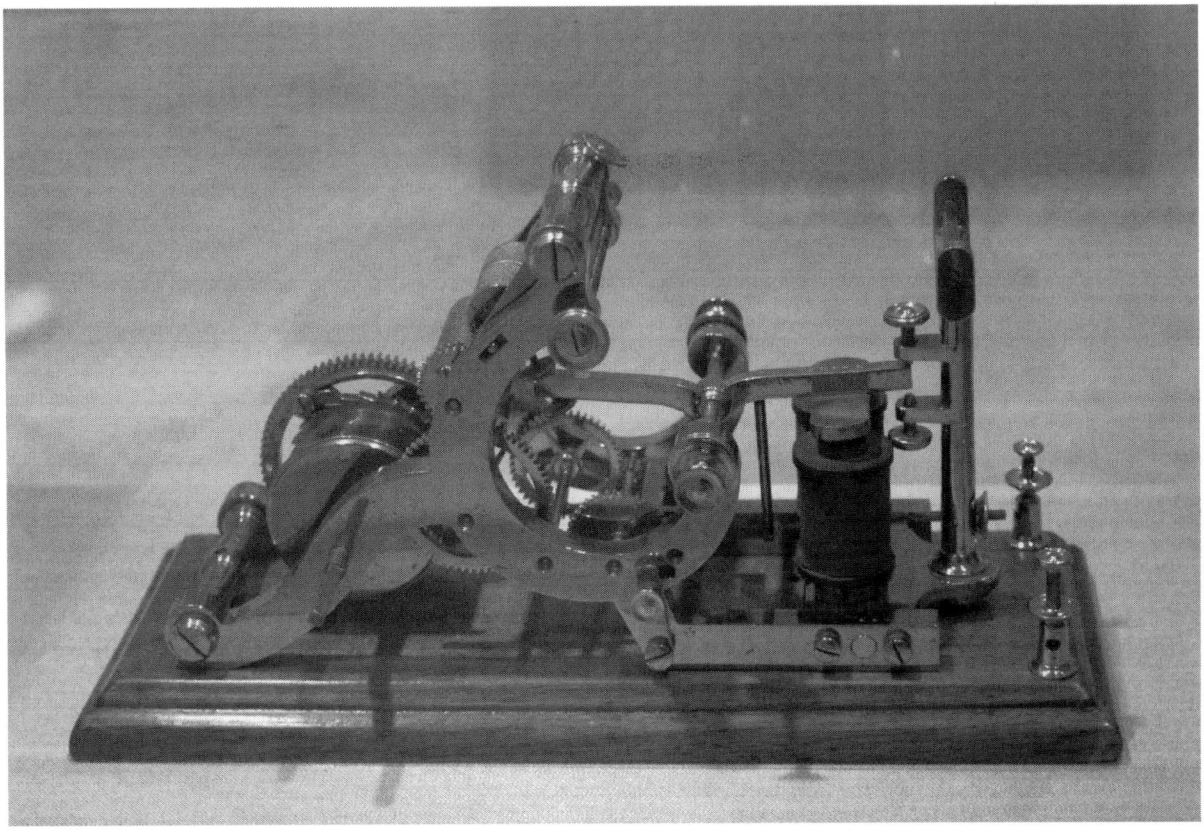

The Logic Behind Morse Code Encoding

Morse code's genius lies in its elegant simplicity, allowing complex messages to be transmitted using just two fundamental elements: dots (·) and dashes (—). Each letter, number, or symbol is assigned a unique sequence of these dots and dashes, enabling efficient encoding and decoding. The logic behind Morse code's encoding system is both intuitive and ingenious, making it a versatile and effective means of communication. In this section, we'll delve into the fundamental principles that govern Morse code's logic.

1. Binary Symbol Representation:
At its core, Morse code operates as a binary system, much like the digital language used in modern computers. However, instead of using 0s and 1s, Morse code employs dots and dashes. This binary approach simplifies transmission and reception, as each character is defined by a specific sequence of these two symbols.

2. Length Distinctions:
The differentiation between dots and dashes isn't arbitrary; it's based on timing. A dot

is a short signal, lasting a single unit of time, while a dash is longer, lasting three units of time. This timing distinction ensures that characters can be easily recognized, even under challenging conditions such as poor signal quality or operator fatigue.

3. Patterns of Characters:
Morse code characters are constructed by combining dots and dashes in unique patterns. The length of the pattern determines the character's identity. Short characters, like the letter "E" (·), consist of just one element, while more complex characters, like "Q" (— — • —), are formed by combining multiple elements.

4. Efficient Encoding:
The allocation of Morse code sequences to characters is based on frequency of use in the English language. More commonly used letters, such as "E" and "T," have shorter Morse code representations, while less frequently used letters have longer codes. This distribution ensures that frequently used characters can be transmitted faster, optimizing communication efficiency.

5. Adaptability:
Morse code's adaptability is one of its defining features. It can be transmitted using various methods, including visual signals (e.g., signal lamps), auditory signals (e.g., sound), or tactile methods (e.g., tapping). This adaptability made it a versatile choice for communication in different contexts, from maritime navigation to military operations.

Understanding the logic behind Morse code encoding is the key to unlocking its power. By grasping the principles of binary representation, timing distinctions, character patterns, efficiency, and adaptability, you'll gain a deeper appreciation for this timeless communication system. As you continue to learn and practice Morse code, you'll uncover its elegance and practicality, making it an invaluable skill for various communication scenarios.

Morse Code Timing and Rhythm (Short and Long Signals)

At the heart of Morse code's efficient communication lies the concept of timing and rhythm. The rhythmic interplay between short signals (dots) and long signals (dashes) allows messages to be encoded and decoded with clarity and precision. Understanding the timing and rhythm of Morse code is essential for both sending and receiving messages accurately. In this section, we'll explore the significance of short and long signals in Morse code, and how they contribute to the system's effectiveness.

1. Short Signals (Dots):
A short signal in Morse code is represented by a single dot (·). It's the fundamental building block of Morse code, and its duration serves as the basic unit of timing. The simplicity of a dot allows for quick transmission and reception, making it a crucial element for encoding and decoding characters efficiently.

2. Long Signals (Dashes):
A long signal in Morse code is represented by a dash (—), which is three times the duration of a dot. This contrast between short and long signals creates a clear distinction that forms the basis for encoding characters. Longer signals introduce a deliberate pause, providing rhythm to the Morse code patterns and allowing recipients to recognize the boundaries between characters and words.

3. Character Timing:
The duration of dots and dashes follows a consistent pattern, ensuring that characters can be distinguished from each other. The timing of dots and dashes, combined with the intercharacter spacing, allows for quick recognition and prevents characters from running together. This timing consistency is critical for effective communication, especially in noisy or challenging environments.

4. Word Spacing:
In addition to the timing of dots and dashes within characters, Morse code also includes spacing between words. This spacing ensures that words within a message are distinct, making it easier for recipients to interpret the message's meaning. Proper word spacing is essential for clarity, and it's achieved using a longer pause than the intercharacter spacing.

By understanding the timing and rhythm of Morse code, you'll gain the ability to send and receive messages accurately. As you practice, you'll develop an intuitive sense of the rhythm, allowing you to recognize characters, words, and punctuation marks with ease. This timing-based system, with its concise dots and dashes, creates a fluid and adaptable language that can be used in various communication scenarios, from maritime operations to radio transmissions, and it remains a valuable skill even in today's digital world.

Morse Code Transmission and Reception Techniques

The effective transmission and reception of Morse code messages require not only a clear understanding of the code's patterns but also the application of specific techniques to ensure accurate communication. Whether you're sending a message via visual signals, auditory signals, or tactile methods, employing the right techniques is essential for successful Morse code communication. In this section, we'll explore various transmission and reception techniques, each tailored to specific scenarios and communication mediums.

1. Visual Transmission and Reception:
Visual transmission of Morse code involves using light signals, such as signal lamps or flashlights, to convey messages. Here are some key techniques for visual Morse code communication:

- **Line of Sight:** Ensure that the transmitting and receiving stations have a clear line of sight to each other, minimizing obstructions and interference.
- **Use of the Morse Code Chart:** Having a visual chart of Morse code characters can be helpful for reference, especially for less experienced operators.
- **Consistent Rhythm:** Maintain a consistent rhythm for dots and dashes, ensuring that the timing remains uniform throughout the transmission.

2. Auditory Transmission and Reception:
Auditory transmission involves sending Morse code messages as sound signals, typically using devices like telegraph keys or sound-producing devices. Here are techniques for effective auditory Morse code communication:

- **Clear Keying:** When using a telegraph key or similar device, focus on clean and distinct keying to produce sharp dots and dashes.
- **Listening Skills:** Develop keen listening skills to differentiate between short signals (dots) and long signals (dashes) and accurately decode the message.
- **Practice with Varied Speeds:** Practice sending and receiving Morse code at different speeds to prepare for various communication scenarios.

3. Tactile Transmission and Reception:
Tactile transmission of Morse code involves using touch or vibration-based methods, making it suitable for scenarios where visual or auditory communication is limited. Here are techniques for effective tactile Morse code communication:

- **Tapping Patterns:** Use consistent tapping patterns for dots and dashes, allowing the recipient to feel the rhythm of the code.
- **Tactile Recognition:** Develop the ability to recognize Morse code characters through touch, whether it's through direct tapping on a surface or using specialized tactile devices.
- **Sensitivity:** Ensure that the recipient is sensitive to tactile signals and can distinguish between different durations (short taps for dots, longer taps for dashes).

4. Communication Protocols:
Establish clear communication protocols, such as starting with "DE" (From) and ending with "AR" (End of Message), to signal the beginning and conclusion of a message. Use common prosigns like "AS" (Wait) to indicate pauses, "KN" (Invitation for a Specific Station to Transmit) for multi-party conversations, and "BT" (Spacing) to separate words clearly.

By mastering these techniques, you'll be well-equipped to send and receive Morse code messages effectively across various communication mediums, ensuring that your transmissions are clear, accurate, and meaningful. Consistent practice and a solid understanding of the fundamentals will enhance your Morse code skills, making you a proficient communicator in this timeless system.

Practice Makes Perfect: Letters and Numbers

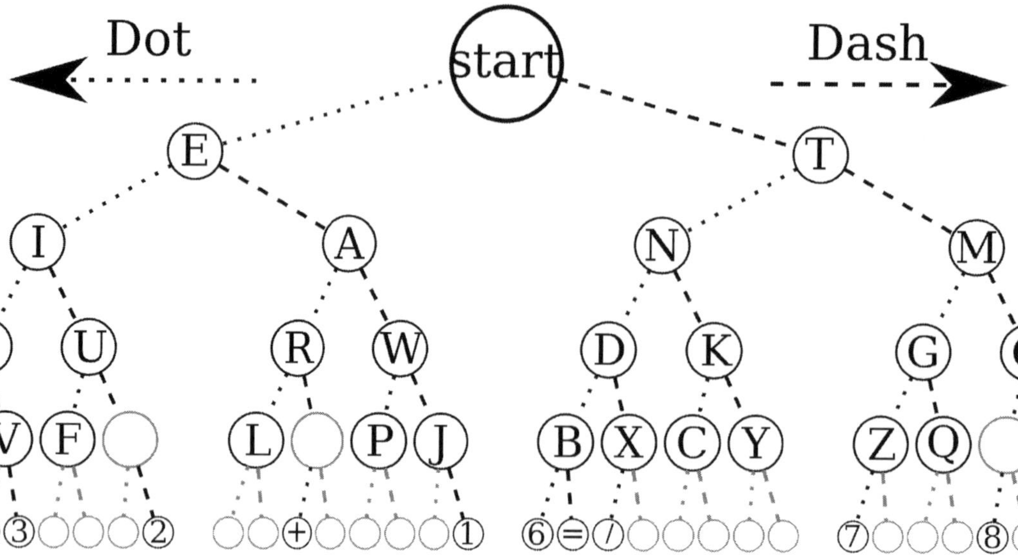

Learning the Morse Code Alphabet Thoroughly

Mastering the Morse code alphabet is the foundation of becoming a proficient Morse code communicator. Each letter, number, and symbol has a unique Morse code representation, and being able to quickly recognize and reproduce these patterns is essential for encoding and decoding messages. In this section, we'll provide you with a systematic approach to learning the Morse code alphabet thoroughly, ensuring that you become confident in using this essential skill.

1. Memorization:
Begin by memorizing the Morse code representations for each letter of the English alphabet, numbers, and common punctuation marks. This initial step is crucial, as it forms the basis for all Morse code communication. Create flashcards, write out the codes, or use mnemonic techniques to help you remember the patterns.

2. Practice with Visual Aids:
Use visual aids, such as charts or diagrams that display the Morse code characters alongside their English equivalents. Seeing the characters side by side helps reinforce the association between the Morse code and the corresponding letters, numbers, or symbols. Regularly review these visual aids to solidify your memory.

3. Practice with Auditory Recognition:
Listen to Morse code transmissions and practice recognizing the characters by sound. There are online resources that provide audio recordings of Morse code, or you can use Morse code software that generates audio signals. Focus on differentiating between dots and dashes, and try to identify the characters as you hear them.

4. Active Reproduction:
Engage in active reproduction of Morse code characters. This involves sending the Morse code for specific letters, numbers, or words using various methods (e.g., tapping, flashing a light, using a telegraph key). By actively reproducing the codes, you reinforce your ability to generate Morse code messages accurately.

5. Word and Phrase Practice:
Once you're comfortable with individual characters, practice forming words and phrases in Morse code. Start with simple words, gradually increasing the complexity as you gain confidence. Use online Morse code translators to convert English text to Morse code and then decode it back to ensure accuracy.

6. Reinforce with Practical Exercises:
Engage in practical exercises that require you to both send and receive Morse code messages. You can find Morse code practice exercises online, or you can practice with a partner who is also learning Morse code. This hands-on experience will help you apply what you've learned and build real-world communication skills.

7. Consistent Review:
Regularly review the Morse code alphabet to prevent forgetting the characters. Periodically test yourself on random characters, and challenge yourself to decode messages quickly. The more consistently you review and practice, the more confident you'll become in your Morse code proficiency.

By following this comprehensive approach to learning the Morse code alphabet, you'll gradually internalize the patterns and become adept at Morse code communication. With practice and dedication, you'll unlock the ability to send and receive messages, opening the door to the rich world of Morse code-based communication.

Practicing Morse Code Transmission and Reception of Letters and Numbers

Effective Morse code communication relies on practice, repetition, and building confidence in your ability to both send and receive messages. In this section, we'll guide you through practical exercises for practicing the transmission and reception of individual letters and numbers. These exercises will help you become fluent in Morse code, enabling you to communicate accurately and efficiently.

Transmission Practice:

To practice transmitting Morse code, you'll need a means of generating dots (·) and dashes (—). You can use various tools for this, such as a telegraph key, a Morse code app, or even your fingers for tapping. Here's how to proceed:

1. Choose a letter or number from the Morse code alphabet.
2. Use your chosen method (key, app, fingers) to send the Morse code for that letter or number.
3. Aim for accuracy in the timing and rhythm of the dots and dashes.
4. Practice sending the same character multiple times to reinforce your muscle memory.

Reception Practice:

To practice receiving Morse code, you'll need a source that generates Morse code signals. This could be an online Morse code generator, an audio recording, or a partner who can transmit the signals. Follow these steps:

1. Have the Morse code signals sent to you (either by an online generator, audio recording, or your practice partner).
2. Listen carefully and pay close attention to the rhythm and timing of the dots and dashes.
3. Try to recognize the Morse code character being transmitted.
4. Practice with both common and less common characters to improve your recognition skills.

Mixed Character Practice:

Combine the transmission and reception exercises to create a more dynamic practice session:

1. Randomly select a letter or number from the Morse code alphabet.
2. First, try to transmit that character accurately (send it yourself).
3. Then, listen to the Morse code being transmitted by an external source (audio, partner) and receive the character.
4. This mixed practice reinforces both your ability to send and receive Morse code, helping you become more proficient.

Word and Number Practice:

Once you're comfortable with individual letters and numbers, progress to practicing entire words or strings of numbers. This exercise simulates real communication scenarios:

1. Choose a word or a string of numbers (e.g., "HELLO" or "12345").
2. Use your chosen method to send the Morse code for the entire word or number sequence.
3. Challenge yourself to decode Morse code messages received from external sources (audio, partner) that contain words or numbers.

By regularly engaging in these practice exercises, you'll build confidence, improve your transmission and reception skills, and develop a natural rhythm for Morse code communication. As you become more proficient, you'll be ready to tackle more complex messages and expand your Morse code capabilities. Remember, practice makes perfect, and the more you practice, the more fluent you'll become in this timeless language of dots and dashes.

Developing Good Morse Code Listening Skills

Listening skills are crucial for proficient Morse code communication. Being able to accurately distinguish between short signals (dots) and long signals (dashes) is essential for receiving messages with clarity. Developing strong Morse code listening skills takes practice, patience, and a systematic approach. In this section, we'll provide you with strategies to enhance your ability to listen effectively to Morse code transmissions.

1. Start with Slow Speeds:

Begin your listening practice at a slower Morse code transmission speed. This allows you to familiarize yourself with the rhythm of dots and dashes and improves your ability to recognize individual characters. As you gain confidence, gradually increase the speed of the transmissions.

2. Focus on Character Distinctions:

Pay close attention to the distinctive patterns that differentiate one character from another. Focus on recognizing the timing of dots and dashes, rather than trying to identify entire words initially. Mastering individual characters is essential before progressing to more complex messages.

3. Practice Regularly:

Consistent practice is key to improving your listening skills. Set aside dedicated time each day to listen to Morse code transmissions. Online resources, Morse code apps, and radio broadcasts (such as amateur radio operators) can provide a steady stream of Morse code for practice.

4. Use Mnemonics:

Consider using mnemonic devices or memory aids to help you remember the Morse code representations for different characters. Create associations between the dots and dashes and something memorable, like a phrase, a visual image, or a personal mnemonic. This can make the learning process more engaging and effective.

5. Decode Simple Words:

Once you're comfortable with individual characters, start practicing with simple words in Morse code. Use online Morse code translators to generate Morse code versions of everyday words. Decode these words and gradually progress to longer sentences as your skills improve.

6. Record and Review:

Record Morse code transmissions that you listen to, and then review them afterward. This allows you to identify areas where you may have misinterpreted characters or missed parts of the message. Reviewing your recordings helps pinpoint areas for improvement.

7. Engage with Other Morse Code Enthusiasts:

Join Morse code forums, online groups, or clubs where enthusiasts gather to exchange Morse code messages. Engaging with other learners and experienced Morse code operators can provide valuable feedback, encouragement, and opportunities for practice.

8. Be Patient and Persistent:

Remember that developing good Morse code listening skills takes time. Be patient with yourself, and celebrate small milestones. Keep practicing, and you'll notice steady improvement over time.

Advanced Morse Code Techniques

Understanding Morse Code Punctuation

In addition to letters and numbers, Morse code includes representations for several common punctuation marks, providing a comprehensive framework for communication. Punctuation marks in Morse code allow for more nuanced and precise messaging, making it possible to convey not only the content but also the tone and structure of a message. In this section, we'll explore the Morse code representations for some key punctuation marks and discuss their significance in Morse code communication.

1. Period (.)

In Morse code, a period is represented by the sequence · — · — · —. It's a simple and recognizable pattern, often used to signal the end of a sentence or a thought. The period provides a clear visual and auditory cue that allows the recipient to anticipate the conclusion of a statement.

2. Comma (,)

The comma in Morse code is represented by the sequence — — • • — —. This symbolizes a brief pause, indicating a separation between elements within a sentence. The comma is a versatile punctuation mark, used to clarify lists, add emphasis, or create a rhythmic flow in the message.

3. Question Mark (?)

A question mark is represented in Morse code as • • — — • •. This pattern serves as a universal indicator of a question, signaling to the recipient that the message requires a response or contains an inquiry. The question mark adds a layer of context, indicating the intended purpose of the message.

4. Exclamation Mark (!)

An exclamation mark is represented by the sequence — • — • — —. In Morse code, it conveys excitement, urgency, or emphasis. Including an exclamation mark in a message ensures that the recipient interprets the tone correctly, adding emotional context to the communication.

5. Ampersand (&)

The ampersand, often used to represent "and," is conveyed in Morse code as • — • • •. This allows for more concise communication when combining words or elements in a message. The ampersand symbolizes the conjunction "and," facilitating the transmission of complex thoughts using fewer characters.

6. Slash (/)

The slash in Morse code is represented by — • • • —. It serves as a separator or indicator of alternatives, helping to segment elements within a message. The slash is useful for conveying choices, options, or divisions in the message's content.

Understanding Morse code punctuation is essential for effectively conveying the intended meaning and structure of your messages. By incorporating these punctuation marks into your Morse code communication, you'll be able to express a wider range of ideas and ensure that your messages are clear, well-structured, and contextually accurate, just like in any written or spoken language.

Prosigns, Q Codes, and Other Special Morse Code Sequences

In addition to the standard Morse code alphabet, a set of specialized sequences, known as prosigns and Q codes, has been developed to streamline communication and convey specific messages more efficiently. These special Morse code sequences have unique meanings, and they play a crucial role in various contexts, from amateur radio operations to maritime navigation. In this section, we'll explore prosigns, Q codes, and other noteworthy special Morse code sequences.

1. Prosigns:

Prosigns, short for procedural signs, are Morse code sequences designed to convey specific procedural or operational messages. They're particularly useful in situations where concise communication is essential. Here are some common prosigns:

- **AR (End of Message):** AR is used to indicate the conclusion of a message. It allows the recipient to know when one complete message ends and another begins.
- **AS (Wait):** AS signals that the sender is pausing temporarily and expects the recipient to wait before responding. This prosign helps manage communication flow.
- **KN (Invitation for a Specific Station to Transmit):** KN directs a specific station to respond or take their turn in a conversation, useful in multi-party communication.
- **BT (Spacing):** BT is used to introduce extra spacing between words or phrases, ensuring message clarity by preventing characters from running together.
- **SK (End of Communication):** SK signals the end of the entire communication session or conversation, providing a more comprehensive conclusion than "End of Message."

2. Q Codes:

Q codes are a set of standardized three-letter abbreviations used in amateur radio operations and other radio communications. They serve as shorthand for common phrases or questions, allowing for efficient exchange of information. Here are a few examples:

- **QTH (What is your location?):** Often used to inquire about a station's geographical location.
- **QSL (I acknowledge receipt):** Used to confirm that a message has been received and understood.
- **QRG (What is my exact frequency?):** Used to request precise frequency information.
- **QRM (Are you being interfered with?):** Used to inquire if a station is experiencing interference from other transmissions.

3. Morse Code Special Sequences:

Apart from prosigns and Q codes, there are other special Morse code sequences with specific meanings, often used in specific contexts. For example:

- **SOS (Distress Signal):** SOS is internationally recognized as a universal distress signal. It's an urgent call for assistance, often used in emergency situations.
- **CQ (Calling All Stations):** CQ is a general call inviting any station that hears the signal to respond. It's used to initiate contact with other stations in amateur radio operations.

These special sequences enhance the efficiency and effectiveness of Morse code communication, providing concise ways to convey important messages, inquiries, or operational instructions. Familiarizing yourself with these prosigns, Q codes, and other special sequences is essential for becoming a skilled Morse code communicator in various contexts.

Tips for Efficient Morse Code Communication

Mastering Morse code takes practice and dedication, but with the right strategies, you can become a skilled communicator in this time-tested system. Whether you're using Morse code for fun, amateur radio, or as a backup means of communication, the following tips will help you improve your efficiency, accuracy, and overall proficiency:

1. Start Slow and Gradually Increase Speed:
When learning Morse code, begin at a slower transmission speed. Focus on mastering individual characters before increasing the speed. As you become more comfortable, gradually raise the speed to challenge yourself and enhance your ability to process Morse code signals at different rates.

2. Maintain a Steady Rhythm:
Consistency in the rhythm of your dots and dashes is crucial for efficient communication. Avoid speeding up during familiar characters or slowing down when encountering less common ones. Practice maintaining a steady pace to improve overall readability.

3. Practice Regularly:
Dedicate consistent time to Morse code practice. Whether it's daily or weekly, regular practice is essential for building and maintaining your skills. Use online resources, join Morse code clubs, or engage with other enthusiasts to keep your Morse code abilities sharp.

4. Focus on Accuracy:
Accuracy is more important than speed. Ensure that each character is transmitted or received correctly. Over time, as your accuracy improves, you'll naturally become faster without sacrificing precision.

5. Learn Common Abbreviations and Prosigns:
Familiarize yourself with common Morse code abbreviations, prosigns, and Q codes. These special sequences allow for more efficient communication and convey specific messages with fewer characters.

6. Listen Actively:
When receiving Morse code, actively listen and focus on the rhythm of the signals. Train your ears to distinguish between dots and dashes, and work on recognizing characters based on their unique patterns.

7. Break Down Words:
Break down longer words or messages into shorter segments. This not only reduces the chance of errors but also helps you manage the flow of information more effectively.

8. Use Standard Procedures:
Adopt standard procedures and protocols for communication. Begin messages with "DE" (From) and end them with "AR" (End of Message) to establish clear communication boundaries.

9. Maintain a Morse Code Journal:
Keep a journal where you record your practice sessions, key insights, challenges, and progress. Reviewing your journal can provide valuable feedback on your strengths and areas for improvement.

10. Be Patient and Persistent:
Learning Morse code is a gradual process. Be patient with yourself, celebrate small achievements, and keep pushing forward. With persistence and dedication, you'll develop efficient Morse code communication skills.

By following these tips, you'll enhance your Morse code abilities and become a confident Morse code communicator. Whether you're transmitting critical information or simply enjoying the art of Morse code, these strategies will help you make the most of this timeless form of communication.

Exploring Morse Code Applications

Morse Code in Emergency Situations

Morse code has a storied history of being a reliable means of communication, especially in emergency situations where conventional communication methods may be unavailable or compromised. Its simplicity, adaptability, and effectiveness make it a valuable tool for sending distress signals, conveying critical information, and facilitating search and rescue efforts. In this section, we'll explore the role of Morse code in emergency scenarios and provide guidance on how to use it effectively when needed.

1. Distress Signals:
One of the most important applications of Morse code in emergencies is sending distress signals. The internationally recognized distress signal is "SOS," which is represented as "... --- ..." (three short signals, three long signals, three short signals). This signal is universally understood as a call for immediate assistance. If you find yourself in a life-threatening situation, transmitting "SOS" in Morse code can alert others that you need help.

2. Visual Signaling:
In situations where visual communication is possible, such as being stranded in an open area or at sea, Morse code can be used with signal lamps, flashlights, mirrors, or other reflective surfaces to transmit messages over long distances. By learning the Morse code alphabet and familiarizing yourself with the SOS distress signal, you can increase your chances of being seen and rescued.

3. Auditory Signaling:
In environments where visual signaling is not feasible, such as during the night or in dense forests, auditory Morse code signals can be effective. Carrying a whistle or using sound-producing devices, you can transmit Morse code signals by short bursts (dots) and longer sounds (dashes) to alert potential rescuers. Transmitting "SOS" audibly can convey the urgency of your situation.

4. Encode Critical Information:
In emergency situations, you may need to communicate critical information, such as your location, the nature of the emergency, or medical conditions. By knowing Morse code, you can encode this information and transmit it more efficiently than using verbal communication, especially if language barriers exist.

5. Responder Communication:
In situations where responders or rescue teams are using Morse code, understanding Morse code can be valuable for receiving instructions, providing updates on your condition, or communicating needs. Being able to understand and respond in Morse code can help coordinate rescue efforts more effectively.

6. Learn Basic Morse Code:
Whether you're an outdoor enthusiast, a boater, or simply concerned about emergency preparedness, learning basic Morse code is a valuable skill. Understanding the Morse code alphabet, the SOS signal, and a few key emergency-related phrases can make a significant difference when communication is critical.

In times of emergency, Morse code can bridge communication gaps and provide a lifeline to safety. By understanding its significance and preparing to use it when needed, you can enhance your emergency preparedness and increase your chances of surviving and receiving assistance in challenging situations.

Morse Code in Amateur Radio (Ham Radio) Operations

Amateur radio, often referred to as "ham radio," is a fascinating and valuable hobby that connects people across the globe through radio waves. Morse code has a significant historical and practical role in amateur radio, and many operators still use it today. Whether for its simplicity, its efficiency, or the thrill of communicating using this classic method, Morse code remains a fundamental part of ham radio. In this section, we'll explore how Morse code is used in amateur radio operations and its importance in this vibrant community.

1. Rich Tradition:
Morse code has a rich tradition in amateur radio. Many seasoned ham radio operators became licensed by demonstrating proficiency in Morse code. As a result, it holds a special place in the hearts of hams worldwide and is a symbol of the history and heritage of the hobby.

2. QRP (Low Power) Operations:
One of the practical applications of Morse code in ham radio is in QRP (low power) operations. Morse code can be effectively transmitted and received using minimal power, making it a popular choice for operators who enjoy the challenge of communicating with limited resources.

3. Long-Distance Communication:
Morse code's ability to be transmitted over long distances with relatively low power makes it valuable for amateur radio operators who want to communicate with other operators around the world. During contests or in challenging propagation conditions, Morse code often remains the most reliable mode for long-distance communication.

4. Emergency Communications:
In emergency situations where standard communication channels may be disrupted, amateur radio operators can use Morse code to relay critical information. Its efficiency in low-power, long-range communication can be a lifeline in emergencies, connecting operators who are providing assistance or coordinating relief efforts.

5. CW (Continuous Wave) Mode:
Morse code is transmitted using continuous wave (CW) mode, which is a simple and

straightforward method of modulation. CW transmissions cut through noise and interference more effectively than some other modes, making it a favorite among operators who enjoy the purity and challenge of this form of communication.

6. Learning and Practice:
For many amateur radio enthusiasts, learning Morse code is a rite of passage. While no longer required for most ham radio licenses, learning Morse code is still a popular pursuit. Ham radio clubs, online communities, and Morse code practice resources allow operators to learn, practice, and refine their skills.

7. Morse Code Contests:
Amateur radio operators often participate in Morse code contests, where they compete to make as many successful contacts as possible within a specified time. These contests provide an exciting opportunity to hone Morse code skills and enjoy friendly competition.

In the ham radio world, Morse code bridges generations, connects enthusiasts worldwide, and serves as a reliable method of communication in diverse scenarios. Whether you're interested in the technical aspects, the historical significance, or the thrill of mastering this classic mode, Morse code continues to play a vital role in the vibrant and diverse community of amateur radio operators.

Morse Code in Modern Technology: Beacon Systems and Beyond

While Morse code has a rich history dating back to the early days of telegraphy, it continues to find applications in modern technology, showcasing its resilience and effectiveness as a communication method. One notable example of Morse code in the modern era is its use in beacon systems, which serve various purposes in navigation, identification, and even artistic expressions. In this section, we'll explore how Morse code is utilized in modern technology and its role in beacon systems, as well as touch on other contemporary applications.

1. Beacon Systems:
Beacon systems are a critical part of modern technology, often used for navigation, signaling, and identification purposes. In some cases, Morse code is integrated into these systems to convey important information efficiently:

- **Nautical Navigation:** Maritime beacons, such as lighthouses and buoys, often use Morse code-like flashing patterns to help ships navigate safely through coastal waters. These visual signals provide valuable information to sailors about their location, the type of hazard, or the characteristics of a particular navigational aid.

- **Aviation:** Some airport beacon systems use Morse code or similar flashing patterns to assist pilots in identifying specific runways, taxiways, or other essential information, especially in low-visibility conditions. These visual cues enhance aviation safety.

- **Radio Beacons:** In radio and communication technology, beacons are used to transmit signals with specific patterns. Morse code can be employed in radio beacons to identify the source or provide additional information about the transmitting station.

2. Emergency Signaling:
While modern communication methods are prevalent, Morse code remains an essential tool for emergency signaling, especially in situations where electronic devices may fail. Morse code's simplicity and low-power requirements make it a viable option for sending distress signals when traditional communication channels are compromised.

3. Artistic Expressions:
In the digital age, Morse code has found a place in artistic and creative endeavors. Some artists and creators use Morse code as a visual or auditory element in their work, combining the classic with the contemporary to convey messages, stories, or themes.

4. Learning and Fun:
While not a primary communication method in most modern technology, learning Morse code remains a popular pursuit. Morse code apps, interactive websites, and educational resources allow individuals to explore this historic communication system for fun, personal development, or as a nod to the past.

5. Innovation and Adaptation:
As technology evolves, new applications of Morse code may emerge. The adaptability of Morse code, its simplicity, and its role in historical and cultural contexts contribute to its potential for innovative uses in the ever-changing landscape of modern technology.

From maritime navigation to aviation safety and from artistic expressions to personal hobbies, Morse code continues to have a place in modern technology. Its enduring relevance highlights its unique qualities and demonstrates how this classic method of communication remains intertwined with our modern world.

Putting It All Together

Real-Life Scenarios for Using Morse Code

While Morse code is no longer the primary means of communication in our digitally connected world, it still finds practical applications in various real-life scenarios. Understanding Morse code can be valuable in situations where traditional communication methods are unavailable, unreliable, or where a covert means of conveying information is necessary. Let's explore some real-life scenarios where Morse code can be useful:

1. Outdoor Adventures:
When exploring remote areas, hiking in the wilderness, or engaging in outdoor activities, you may find yourself in situations where cell phone signals are weak or nonexistent. Morse code can be an effective way to communicate with fellow adventurers, signal for help, or convey important messages to others in your group.

2. Emergency Situations:
In emergencies where modern communication devices may fail due to power outages, network disruptions, or damaged infrastructure, Morse code can serve as a reliable backup method. Sending distress signals using Morse code's simple visual or auditory signals can help rescuers locate you or alert others to your situation.

3. Maritime Navigation:
Morse code-like flashing patterns are still used in maritime navigation aids, such as lighthouses, buoys, and navigational beacons. Understanding these patterns can assist sailors and boaters in identifying their position, the type of hazard, or the characteristics of a particular navigational aid, enhancing safety on the water.

4. Aviation Safety:
In aviation, Morse code or similar visual patterns are used in some beacon systems at airports. Pilots can use these cues, especially in low-visibility conditions, to identify specific runways, taxiways, or other essential information, contributing to safer landings and takeoffs.

5. Amateur Radio (Ham Radio) Operations:
Morse code remains a beloved aspect of the amateur radio hobby. Many ham radio operators use Morse code to communicate, especially during contests or in situations where long-distance communication is challenging. If you're part of the amateur radio community, Morse code proficiency is essential for effective communication.

6. Covert Communication:
In scenarios where you need to convey information without drawing attention, Morse code can be a discreet means of communication. It allows you to transmit messages without speaking or using visible gestures, making it useful for covert operations or situations where secrecy is crucial.

7. Historical Reenactments and Traditions:
In historical reenactments, particularly those involving military or maritime themes, participants often use Morse code to recreate the communication methods of the past. Additionally, some organizations and clubs preserve Morse code traditions, using it for special events or commemorations.

8. Personal Growth and Learning:
For many enthusiasts, learning Morse code is a rewarding personal challenge. Whether you're interested in historical communication, enjoy exploring unique skills, or simply want to expand your knowledge, learning Morse code can be a fulfilling endeavor.

While Morse code may not be part of our everyday communication, its versatility and historical significance make it a valuable tool in specific situations. By understanding and practicing Morse code, you can be prepared for unexpected scenarios, enhance your outdoor experiences, and appreciate this timeless form of communication.

Complex Message Encoding and Decoding

Encoding and decoding complex messages in Morse code involves applying the foundational knowledge of the Morse code alphabet, timing, rhythm, and special sequences to convey more extensive and nuanced information. This advanced skill allows you to transmit detailed messages, including sentences, paragraphs, or even entire conversations. In this section, we'll explore the techniques and considerations for encoding and decoding complex messages in Morse code.

1. Segment Your Message:
When encoding a complex message, it's essential to break it down into manageable segments. Divide the message into words, phrases, or logical units that can be transmitted and decoded individually. This segmentation prevents confusion and ensures the accuracy of the transmission.

2. Use Punctuation and Prosigns:
Incorporate punctuation marks, prosigns (e.g., AR for "End of Message," BT for spacing), and Q codes where appropriate. Punctuation helps structure the message, convey tone, and clarify the intended meaning. Prosigns and Q codes allow for efficient communication of common phrases and instructions.

3. Maintain Clarity and Timing:
Keep the Morse code transmission clear and well-timed. Maintain consistent spacing between characters and words, ensuring that the receiver can distinguish between letters, numbers, and punctuation. Clear timing prevents characters from running together, enhancing readability.

4. Practice Comprehensive Vocabulary:
Expand your Morse code vocabulary to include common words, names, and terms you're likely to encounter in your messages. Practice encoding and decoding these words to improve your efficiency and confidence in transmitting complex messages.

5. Practice Decoding:
Decoding complex messages requires active listening skills and familiarity with Morse code patterns. Practice decoding Morse code transmissions of varying lengths and complexity to enhance your ability to understand messages accurately and quickly.

6. Use Digital Resources:
Leverage digital tools, such as Morse code translator apps or online resources, to help you encode longer messages and decode unfamiliar terms. These tools can be valuable for practicing, checking your accuracy, and expanding your Morse code vocabulary.

7. Focus on Context:
Consider the context of the message you're encoding or decoding. The surrounding information or the purpose of the communication can provide clues that aid in understanding the intended meaning of the message, especially when dealing with longer or more intricate messages.

8. Maintain Patience and Persistence:
Encoding and decoding complex messages in Morse code require practice and patience. Be persistent in refining your skills, reviewing Morse code rules, and consistently expanding your vocabulary. Over time, your proficiency will improve, allowing you to handle more complex communication scenarios.

By mastering complex message encoding and decoding, you'll be prepared to effectively communicate detailed information, convey messages with precision, and fully appreciate the depth and versatility of Morse code. This skill opens up new possibilities for meaningful conversations and demonstrates the enduring relevance of Morse code in the modern world.

Building Confidence in Morse Code Communication

Confidence in Morse code communication is essential for effective transmission and reception of messages. Whether you're a beginner learning the basics or an experienced operator looking to enhance your skills, building confidence can significantly impact your ability to use Morse code comfortably and accurately. In this section, we'll explore strategies to boost your confidence in Morse code communication.

1. Master the Fundamentals:
A solid foundation is crucial. Ensure that you have a thorough understanding of the Morse code alphabet, timing, and the basics of encoding and decoding. Practice individual characters until you can recognize and send them without hesitation.

2. Practice Regularly:
Consistent practice is the key to building confidence. Dedicate regular time to Morse code practice, whether it's daily drills, practice sessions with online resources, or participation in Morse code events or contests.

3. Set Achievable Goals:
Establish achievable goals based on your current skill level. Start with short messages, and gradually work your way up to more complex communication. Celebrate each milestone, and recognize the progress you're making.
4. Engage with the Community:
Join Morse code clubs, online forums, or social media groups dedicated to Morse code enthusiasts. Engaging with a community of like-minded individuals can provide support, motivation, and opportunities to exchange tips and experiences.

5. Embrace Mistakes:
Don't be discouraged by mistakes. They're a natural part of the learning process. Use mistakes as learning opportunities, identify areas that need improvement, and strive to do better next time.

6. Focus on Fluency, Not Speed:
While speed can be a goal, prioritize fluency and accuracy over velocity. Develop a natural rhythm in sending and receiving Morse code. As you become more fluent, speed will naturally improve.

7. Personalize the Learning Experience:
Customize your practice to suit your interests and goals. If you're passionate about certain topics, encode messages related to those topics. Tailoring the learning experience to your preferences can make it more engaging.

8. Learn from Successful Transmissions:
When you successfully transmit or decode a message, take note of what went well. Analyze your strengths during those moments, and replicate those practices in future communications.

9. Stay Positive and Patient:
Building confidence takes time. Stay positive, be patient with yourself, and maintain a growth mindset. Each practice session is a step forward on your journey to becoming a skilled Morse code communicator.

10. Step Out of Your Comfort Zone:
Challenge yourself by participating in Morse code activities that push you out of your comfort zone. Try sending longer messages, experimenting with different speeds, or engaging in conversations with fellow Morse code enthusiasts.

By implementing these strategies and embracing a gradual, persistent approach, you'll build confidence in Morse code communication. Confidence not only enhances your skills but also allows you to fully enjoy the practicality of this timeless system.

Practice Exercises

Exercise 1.1 : Practice writing the Morse code letter for each English letter of the alphabet.

Fill in the boxes next to each letter with the Morse Code form 5 times.

A	.−				
B					
C					
D					
E					

F					
G					
H					
I					
J					
K					
L					

M					
N					
O					
P					
Q					
R					
S					

T					
U					
V					
W					
X					
Y					
Z					

Exercise 1.2 : Decode the provided Morse code letters to identify the English letters.

Write the English letter in the boxes provided below each Morse Code form:

-...	.--.	--..	-..-	..	...-
--.-	--	....	..-.	..-	.---
-	...	.-..	.--	.-	-.-
-..	--.	-.-.	---	-.	.
-.--	.-.				

Exercise 2.1 : Translate the following English words into Morse code.

H E L L O	....	.	.-..	.-..	---
W O R L D					
M O R S E					
L E A R N					
T I G E R					
A P P L E					
T A B L E					
Q U I E T					
C L O C K					

F R O G S					
S H A R K					
W A V E S					
J U M P S					
G R A S S					
C H A I R					
M O U N T					
S T O R M					
B I R D S					
F L O O R					

R O B O T ☐ ☐ ☐ ☐ ☐

W A T E R ☐ ☐ ☐ ☐ ☐

T H O R N ☐ ☐ ☐ ☐ ☐

L A U G H ☐ ☐ ☐ ☐ ☐

S M O K E ☐ ☐ ☐ ☐ ☐

B R I C K ☐ ☐ ☐ ☐ ☐

P L A N T ☐ ☐ ☐ ☐ ☐

C L O U D ☐ ☐ ☐ ☐ ☐

S T A R S ☐ ☐ ☐ ☐ ☐

F R U I T ☐ ☐ ☐ ☐ ☐

Exercise 2.2 : Write the Morse code for the phrase.

Write the Morse Code that corresponds to each English phrase on the lines below.

I LOVE MORSE CODE

.. / .-.. --- ...- . / -- ---

.-. ... . / -.-. --- -.. .

LIFE IS BEAUTIFUL

DREAM BIG, WORK HARD

FOLLOW YOUR HEART

POSITIVITY MATTERS

NEVER STOP LEARNING

BE KIND TO OTHERS

PERSISTENCE PAYS OFF

SPREAD LOVE AND JOY

AIM HIGH, ACHIEVE MUCH

EMBRACE THE JOURNEY

SEEK ADVENTURES

LIVE WITH PURPOSE

EXPRESS YOURSELF

BE BRAVE, STAY TRUE

Exercise 3: Translate the following Morse code sequences back into English words.

-.. .-. . .- -- ... / -.. --- / -.-.
--- -- . / -.- --- -- . / - - .-. ..- .

DREAMS DO

COME TRUE

-... . .-.. .. . ...- . / .. -. / -.--
--- ..- .-. / -... -.-- .. -. ..-. - .-
.- .-.. ...-

-.-. .... .- .-.. .-.. . -. --. . ... /
-- .- -.- . / ..-. ..- .- .-.. -.- ... /
-- .- -.-

-. . ...- . .-. / --. .. ...- . / ..-
.--. / .... --- .--. .

..-. --- .-.. .-.. --- .-- / -.-- ---
..- .-. / .... .- .-. -

.- -.-. ....- . / -.-- --- ..-
.-. / -.. .-. . .- -- ...

... . . -.- / - .... . / .- -.. ...- .
-. - .. - .- .-. -.

. -. .--- --- -.-- / - .... . / .---
--- ..- .-. -. . -. -.--

. -- -... .-. .- -.-. . / - .... . /
..-. -. -... .-. -.- --- -. .-- -.- -.
.- -.

. -..- .--. .-.. --- .-. . / -. . .-- /
.... --- .-. .. --.. --- -. ...

..-. --- .-.. .-.. --- .-- / -.-- ---
..- .-. / -.. .-. . .- -- ...

.--. ..- .-. ... ..- . / -.-- --- ..- .-. / .--. .- ... . --- -.

.-.. .. ...- . / -.-- --- ..- .-. / -... . ... - / .-.. .. ..- .

-.-. .... . .-. .. .. / . ...- ...-. -.-- / -- --- -- . - -. -

-.-. .-. . .- - / -.-- --- ..- .-. / --- .-- -. / .-- --- -. / .--. .- - .- ....

. -. .--- --- -.-- / . ...- ...-. -.--
/ -- --- -- . -. -

-.-. .-. . .- - / -.-- --- ..- .-. /
..-. ..- - ..- .-. .

-.. .-. . .- -- ... / -... . -.-. ---
-.-- .- -- . / .-. . .- .-.. .. - -.--

.-.. --- ...- . / -.- --- -. --. ..- .-.
... / .- .-.. .-..

.-- --- .-. -.- / .... .- .-. -.. /
.--. .- .-.. -.-- / .... .- .-. -..

... -- .. .-.. . / . ...- ...-. .-. -.--
-.. .- -.--

. -..- .--. .-. . ... / -.-- --- ..-
.-. ... . -.-.

. .-.. .- .-. -. / ..-. .-. --- -- /
.... - --- .-. -.--

... - .- -.-- / - .-. ..- . / - --- / -.-- --- ..-

.- -.. ...- . -. - ..- .-. . / .- .-- .- .. - ...

-... . .-.. .. . ...- . / .- -. -.. / .- -.-. .... - ... .

.-.. .. ...- . / .-- .. - .... / .-- .. - .... / .--. .- ... . --- -.

-. .- - ..- .-. . ... / ' ... -... . .-
..- - -.--

--. .-. .- - .. - .. - ..- - .- / .. ...
/ -.- . -.--

.. -. ... .--. .. .-. . / .--. --- ...
.. - .. ..- - . - -.--

.-.. --- ...- . / .- -. -.. / -.- .. -.
-.. -. ... ...

Dear Reader,

If you've enjoyed delving into the world of Morse code with "Learn Morse Code: A Step-By-Step Manual and Workbook for Beginners," I kindly invite you to consider leaving a glowing 5-star review on Amazon. Your positive feedback fuels the author, Jonathan Harrison, to continue crafting valuable content that enriches your learning experience and keeps you engaged.

By sharing your thoughts and awarding the book with 5 stars, you not only encourage the author but also help other curious minds discover this treasure of Morse code knowledge. The more support we receive, the more inspired Jonathan becomes to create more informative and exciting materials to cater to your interests.

So, please take a moment to show your appreciation with a 5-star review on Amazon. Your feedback truly matters, and it's an excellent way to ensure we keep delivering quality content that makes learning both enjoyable and rewarding. Thank you for being a part of this journey!

Warm Regards,

Jonathan Harrison

Blank Practice Pages

A: • —	E: •	I: • •	M: — —	Q: — — • —	U: • • —	
B: — • • •	F: • • — •	J: • — — —	N: — •	R: • — •	V: • • • —	Y: — • — —
C: — • — •	G: — — •	K: — • —	O: — — —	S: • • •	W: • — —	Z: — — • •
D: — • •	H: • • • •	L: • — • •	P: • — — •	T: —	X: — • • —	

A: • —	E: •	I: • •	M: — —	Q: — — • —	U: • • —	
B: — • • •	F: • • — •	J: • — — —	N: — •	R: • — •	V: • • • —	Y: — • — —
C: — • — •	G: — — •	K: — • —	O: — — —	S: • • •	W: • — —	Z: — — • •
D: — • •	H: • • • •	L: • — • •	P: • — — •	T: —	X: — • • —	

A: • —	E: •	I: • •	M: — —	Q: — — • —	U: • • —	
B: — • • •	F: • • — •	J: • — — —	N: — •	R: • — •	V: • • • —	Y: — • — —
C: — • — •	G: — — •	K: — • —	O: — — —	S: • • •	W: • — —	Z: — — • •
D: — • •	H: • • • •	L: • — • •	P: • — — •	T: —	X: — • • —	

A: • —	E: •	I: • •	M: — —	Q: — — • —	U: • • —	
B: — • • •	F: • • — •	J: • — — —	N: — •	R: • — •	V: • • • —	Y: — • — —
C: — • — •	G: — — •	K: — • —	O: — — —	S: • • •	W: • — —	Z: — — • •
D: — • •	H: • • • •	L: • — • •	P: • — — •	T: —	X: — • • —	

A: • —	E: •	I: • •	M: — —	Q: — — • —	U: • • —	
B: — • • •	F: • • — •	J: • — — —	N: — •	R: • — •	V: • • • —	Y: — • — —
C: — • — •	G: — — •	K: — • —	O: — — —	S: • • •	W: • — —	Z: — — • •
D: — • •	H: • • • •	L: • — • •	P: • — — •	T: —	X: — • • —	

A: • —	E: •	I: • •	M: — —	Q: — — • —	U: • • —	
B: — • • •	F: • • — •	J: • — — —	N: — •	R: • — •	V: • • • —	Y: — • — —
C: — • — •	G: — — •	K: — • —	O: — — —	S: • • •	W: • — —	Z: — — • •
D: — • •	H: • • • •	L: • — • •	P: • — — •	T: —	X: — • • —	

A: • —	E: •	I: • •	M: — —	Q: — — • —	U: • • —	
B: — • • •	F: • • — •	J: • — — —	N: — •	R: • — •	V: • • • —	Y: — • — —
C: — • — •	G: — — •	K: — • —	O: — — —	S: • • •	W: • — —	Z: — — • •
D: — • •	H: • • • •	L: • — • •	P: • — — •	T: —	X: — • • —	

A: • —	E: •	I: • •	M: — —	Q: — — • —	U: • • —	
B: — • • •	F: • • — •	J: • — — —	N: — •	R: • — •	V: • • • —	Y: — • — —
C: — • — •	G: — — •	K: — • —	O: — — —	S: • • •	W: • — —	Z: — — • •
D: — • •	H: • • • •	L: • — • •	P: • — — •	T: —	X: — • • —	

A: • —	E: •	I: • •	M: — —	Q: — — • —	U: • • —	
B: — • • •	F: • • — •	J: • — — —	N: — •	R: • — •	V: • • • —	Y: — • — —
C: — • — •	G: — — •	K: — • —	O: — — —	S: • • •	W: • — —	Z: — — • •
D: — • •	H: • • • •	L: • — • •	P: • — — •	T: —	X: — • • —	

A: • —	E: •	I: • •	M: — —	Q: — — • —	U: • • —	
B: — • • •	F: • • — •	J: • — — —	N: — •	R: • — •	V: • • • —	Y: — • — —
C: — • — •	G: — — •	K: — • —	O: — — —	S: • • •	W: • — —	Z: — — • •
D: — • •	H: • • • •	L: • — • •	P: • — — •	T: —	X: — • • —	

A: • —	E: •	I: • •	M: — —	Q: — — • —	U: • • —	
B: — • • •	F: • • — •	J: • — — —	N: — •	R: • — •	V: • • • —	Y: — • — —
C: — • — •	G: — — •	K: — • —	O: — — —	S: • • •	W: • — —	Z: — — • •
D: — • •	H: • • • •	L: • — • •	P: • — — •	T: —	X: — • • —	

A: • —	E: •	I: • •	M: — —	Q: — — • —	U: • • —	
B: — • • •	F: • • — •	J: • — — —	N: — •	R: • — •	V: • • • —	Y: — • — —
C: — • — •	G: — — •	K: — • —	O: — — —	S: • • •	W: • — —	Z: — — • •
D: — • •	H: • • • •	L: • — • •	P: • — — •	T: —	X: — • • —	

A: • —	E: •	I: • •	M: — —	Q: — — • —	U: • • —	
B: — • • •	F: • • — •	J: • — — —	N: — •	R: • — •	V: • • • —	Y: — • — —
C: — • — •	G: — — •	K: — • —	O: — — —	S: • • •	W: • — —	Z: — — • •
D: — • •	H: • • • •	L: • — • •	P: • — — •	T: —	X: — • • —	

A: • —	E: •	I: • •	M: — —	Q: — — • —	U: • • —	
B: — • • •	F: • • — •	J: • — — —	N: — •	R: • — •	V: • • • —	Y: — • — —
C: — • — •	G: — — •	K: — • —	O: — — —	S: • • •	W: • — —	Z: — — • •
D: — • •	H: • • • •	L: • — • •	P: • — — •	T: —	X: — • • —	

A: • —	E: •	I: • •	M: — —	Q: — — • —	U: • • —	
B: — • • •	F: • • — •	J: • — — —	N: — •	R: • — •	V: • • • —	Y: — • — —
C: — • — •	G: — — •	K: — • —	O: — — —	S: • • •	W: • — —	Z: — — • •
D: — • •	H: • • • •	L: • — • •	P: • — — •	T: —	X: — • • —	

A: • —	E: •	I: • •	M: — —	Q: — — • —	U: • • —	
B: — • • •	F: • • — •	J: • — — —	N: — •	R: • — •	V: • • • —	Y: — • — —
C: — • — •	G: — — •	K: — • —	O: — — —	S: • • •	W: • — —	Z: — — • •
D: — • •	H: • • • •	L: • — • •	P: • — — •	T: —	X: — • • —	

A: • —	E: •	I: • •	M: — —	Q: — — • —	U: • • —	
B: — • • •	F: • • — •	J: • — — —	N: — •	R: • — •	V: • • • —	Y: — • — —
C: — • — •	G: — — •	K: — • —	O: — — —	S: • • •	W: • — —	Z: — — • •
D: — • •	H: • • • •	L: • — • •	P: • — — •	T: —	X: — • • —	

A: • —	E: •	I: • •	M: — —	Q: — — • —	U: • • —	
B: — • • •	F: • • — •	J: • — — —	N: — •	R: • — •	V: • • • —	Y: — • — —
C: — • — •	G: — — •	K: — • —	O: — — —	S: • • •	W: • — —	Z: — — • •
D: — • •	H: • • • •	L: • — • •	P: • — — •	T: —	X: — • • —	

A: • —	E: •	I: • •	M: — —	Q: — — • —	U: • • —	
B: — • • •	F: • • — •	J: • — — —	N: — •	R: • — •	V: • • • —	Y: — • — —
C: — • — •	G: — — •	K: — • —	O: — — —	S: • • •	W: • — —	Z: — — • •
D: — • •	H: • • • •	L: • — • •	P: • — — •	T: —	X: — • • —	

A: • —	E: •	I: • •	M: — —	Q: — — • —	U: • • —	
B: — • • •	F: • • — •	J: • — — —	N: — •	R: • — •	V: • • • —	Y: — • — —
C: — • — •	G: — — •	K: — • —	O: — — —	S: • • •	W: • — —	Z: — — • •
D: — • •	H: • • • •	L: • — • •	P: • — — •	T: —	X: — • • —	

A: • —	E: •	I: • •	M: — —	Q: — — • —	U: • • —	
B: — • • •	F: • • — •	J: • — — —	N: — •	R: • — •	V: • • • —	Y: — • — —
C: — • — •	G: — — •	K: — • —	O: — — —	S: • • •	W: • — —	Z: — — • •
D: — • •	H: • • • •	L: • — • •	P: • — — •	T: —	X: — • • —	

A: • —	E: •	I: • •	M: — —	Q: — — • —	U: • • —	
B: — • • •	F: • • — •	J: • — — —	N: — •	R: • — •	V: • • • —	Y: — • — —
C: — • — •	G: — — •	K: — • —	O: — — —	S: • • •	W: • — —	Z: — — • •
D: — • •	H: • • • •	L: • — • •	P: • — — •	T: —	X: — • • —	

A: · —	E: ·	I: · ·	M: — —	Q: — — · —	U: · · —	
B: — · · ·	F: · · — ·	J: · — — —	N: — ·	R: · — ·	V: · · · —	Y: — · — —
C: — · — ·	G: — — ·	K: — · —	O: — — —	S: · · ·	W: · — —	Z: — — · ·
D: — · ·	H: · · · ·	L: · — · ·	P: · — — ·	T: —	X: — · · —	

A: • —	E: •	I: • •	M: — —	Q: — — • —	U: • • —	
B: — • • •	F: • • — •	J: • — — —	N: — •	R: • — •	V: • • • —	Y: — • — —
C: — • — •	G: — — •	K: — • —	O: — — —	S: • • •	W: • — —	Z: — — • •
D: — • •	H: • • • •	L: • — • •	P: • — — •	T: —	X: — • • —	

A: • —	E: •	I: • •	M: — —	Q: — — • —	U: • • —	
B: — • • •	F: • • — •	J: • — — —	N: — •	R: • — •	V: • • • —	Y: — • — —
C: — • — •	G: — — •	K: — • —	O: — — —	S: • • •	W: • — —	Z: — — • •
D: — • •	H: • • • •	L: • — • •	P: • — — •	T: —	X: — • • —	

A: • —	E: •	I: • •	M: — —	Q: — — • —	U: • • —	
B: — • • •	F: • • — •	J: • — — —	N: — •	R: • — •	V: • • • —	Y: — • — —
C: — • — •	G: — — •	K: — • —	O: — — —	S: • • •	W: • — —	Z: — — • •
D: — • •	H: • • • •	L: • — • •	P: • — — •	T: —	X: — • • —	

A: • —	E: •	I: • •	M: — —	Q: — — • —	U: • • —	
B: — • • •	F: • • — •	J: • — — —	N: — •	R: • — •	V: • • • —	Y: — • — —
C: — • — •	G: — — •	K: — • —	O: — — —	S: • • •	W: • — —	Z: — — • •
D: — • •	H: • • • •	L: • — • •	P: • — — •	T: —	X: — • • —	

A: • —	E: •	I: • •	M: — —	Q: — — • —	U: • • —	
B: — • • •	F: • • — •	J: • — — —	N: — •	R: • — •	V: • • • —	Y: — • — —
C: — • — •	G: — — •	K: — • —	O: — — —	S: • • •	W: • — —	Z: — — • •
D: — • •	H: • • • •	L: • — • •	P: • — — •	T: —	X: — • • —	

A: • —	E: •	I: • •	M: — —	Q: — — • —	U: • • —	
B: — • • •	F: • • — •	J: • — — —	N: — •	R: • — •	V: • • • —	Y: — • — —
C: — • — •	G: — — •	K: — • —	O: — — —	S: • • •	W: • — —	Z: — — • •
D: — • •	H: • • • •	L: • — • •	P: • — — •	T: —	X: — • • —	

A: · —	E: ·	I: · ·	M: — —	Q: — — · —	U: · · —	
B: — · · ·	F: · · — ·	J: · — — —	N: — ·	R: · — ·	V: · · · —	Y: — · — —
C: — · — ·	G: — — ·	K: — · —	O: — — —	S: · · ·	W: · — —	Z: — — · ·
D: — · ·	H: · · · ·	L: · — · ·	P: · — — ·	T: —	X: — · · —	

A: • —	E: •	I: • •	M: — —	Q: — — • —	U: • • —	
B: — • • •	F: • • — •	J: • — — —	N: — •	R: • — •	V: • • • —	Y: — • — —
C: — • — •	G: — — •	K: — • —	O: — — —	S: • • •	W: • — —	Z: — — • •
D: — • •	H: • • • •	L: • — • •	P: • — — •	T: —	X: — • • —	

A: • —	E: •	I: • •	M: — —	Q: — — • —	U: • • —	
B: — • • •	F: • • — •	J: • — — —	N: — •	R: • — •	V: • • • —	Y: — • — —
C: — • — •	G: — — •	K: — • —	O: — — —	S: • • •	W: • — —	Z: — — • •
D: — • •	H: • • • •	L: • — • •	P: • — — •	T: —	X: — • • —	

A: • —	E: •	I: • •	M: — —	Q: — — • —	U: • • —	
B: — • • •	F: • • — •	J: • — — —	N: — •	R: • — •	V: • • • —	Y: — • — —
C: — • — •	G: — — •	K: — • —	O: — — —	S: • • •	W: • — —	Z: — — • •
D: — • •	H: • • • •	L: • — • •	P: • — — •	T: —	X: — • • —	

A: • —	E: •	I: • •	M: — —	Q: — — • —	U: • • —	
B: — • • •	F: • • — •	J: • — — —	N: — •	R: • — •	V: • • • —	Y: — • — —
C: — • — •	G: — — •	K: — • —	O: — — —	S: • • •	W: • — —	Z: — — • •
D: — • •	H: • • • •	L: • — • •	P: • — — •	T: —	X: — • • —	

A: • —	E: •	I: • •	M: — —	Q: — — • —	U: • • —	
B: — • • •	F: • • — •	J: • — — —	N: — •	R: • — •	V: • • • —	Y: — • — —
C: — • — •	G: — — •	K: — • —	O: — — —	S: • • •	W: • — —	Z: — — • •
D: — • •	H: • • • •	L: • — • •	P: • — — •	T: —	X: — • • —	

A: • —	E: •	I: • •	M: — —	Q: — — • —	U: • • —	
B: — • • •	F: • • — •	J: • — — —	N: — •	R: • — •	V: • • • —	Y: — • — —
C: — • — •	G: — — •	K: — • —	O: — — —	S: • • •	W: • — —	Z: — — • •
D: — • •	H: • • • •	L: • — • •	P: • — — •	T: —	X: — • • —	

A: • —	E: •	I: • •	M: — —	Q: — — • —	U: • • —	
B: — • • •	F: • • — •	J: • — — —	N: — •	R: • — •	V: • • • —	Y: — • — —
C: — • — •	G: — — •	K: — • —	O: — — —	S: • • •	W: • — —	Z: — — • •
D: — • •	H: • • • •	L: • — • •	P: • — — •	T: —	X: — • • —	

A: • —	E: •	I: • •	M: — —	Q: — — • —	U: • • —	
B: — • • •	F: • • — •	J: • — — —	N: — •	R: • — •	V: • • • —	Y: — • — —
C: — • — •	G: — — •	K: — • —	O: — — —	S: • • •	W: • — —	Z: — — • •
D: — • •	H: • • • •	L: • — • •	P: • — — •	T: —	X: — • • —	

A: • —	E: •	I: • •	M: — —	Q: — — • —	U: • • —	
B: — • • •	F: • • — •	J: • — — —	N: — •	R: • — •	V: • • • —	Y: — • — —
C: — • — •	G: — — •	K: — • —	O: — — —	S: • • •	W: • — —	Z: — — • •
D: — • •	H: • • • •	L: • — • •	P: • — — •	T: —	X: — • • —	

A: · —	E: ·	I: · ·	M: — —	Q: — — · —	U: · · —	
B: — · · ·	F: · · — ·	J: · — — —	N: — ·	R: · — ·	V: · · · —	Y: — · — —
C: — · — ·	G: — — ·	K: — · —	O: — — —	S: · · ·	W: · — —	Z: — — · ·
D: — · ·	H: · · · ·	L: · — · ·	P: · — — ·	T: —	X: — · · —	

A: • —	E: •	I: • •	M: — —	Q: — — • —	U: • • —	
B: — • • •	F: • • — •	J: • — — —	N: — •	R: • — •	V: • • • —	Y: — • — —
C: — • — •	G: — — •	K: — • —	O: — — —	S: • • •	W: • — —	Z: — — • •
D: — • •	H: • • • •	L: • — • •	P: • — — •	T: —	X: — • • —	

A: • —	E: •	I: • •	M: — —	Q: — — • —	U: • • —	
B: — • • •	F: • • — •	J: • — — —	N: — •	R: • — •	V: • • • —	Y: — • — —
C: — • — •	G: — — •	K: — • —	O: — — —	S: • • •	W: • — —	Z: — — • •
D: — • •	H: • • • •	L: • — • •	P: • — — •	T: —	X: — • • —	

A: • —	E: •	I: • •	M: — —	Q: — — • —	U: • • —	
B: — • • •	F: • • — •	J: • — — —	N: — •	R: • — •	V: • • • —	Y: — • — —
C: — • — •	G: — — •	K: — • —	O: — — —	S: • • •	W: • — —	Z: — — • •
D: — • •	H: • • • •	L: • — • •	P: • — — •	T: —	X: — • • —	

A: • —	E: •	I: • •	M: — —	Q: — — • —	U: • • —	
B: — • • •	F: • • — •	J: • — — —	N: — •	R: • — •	V: • • • —	Y: — • — —
C: — • — •	G: — — •	K: — • —	O: — — —	S: • • •	W: • — —	Z: — — • •
D: — • •	H: • • • •	L: • — • •	P: • — — •	T: —	X: — • • —	

A: · —	E: ·	I: · ·	M: — —	Q: — — · —	U: · · —	
B: — · · ·	F: · · — ·	J: · — — —	N: — ·	R: · — ·	V: · · · —	Y: — · — —
C: — · — ·	G: — — ·	K: — · —	O: — — —	S: · · ·	W: · — —	Z: — — · ·
D: — · ·	H: · · · ·	L: · — · ·	P: · — — ·	T: —	X: — · · —	

A: • —	E: •	I: • •	M: — —	Q: — — • —	U: • • —	
B: — • • •	F: • • — •	J: • — — —	N: — •	R: • — •	V: • • • —	Y: — • — —
C: — • — •	G: — — •	K: — • —	O: — — —	S: • • •	W: • — —	Z: — — • •
D: — • •	H: • • • •	L: • — • •	P: • — — •	T: —	X: — • • —	

Printed in Great Britain
by Amazon

34199085R00062